Programming of Life Prerequisites

Physical Constants and Properties Requirements

By Donald E Johnson
©2011

Published by Big Mac Publishers October, 2011
www.bigmacpublishers.com Sylacauga, AL
Printed and bound in the USA

Programming of Life Prerequisites

Physical Constants and Properties Requirements

By Donald E Johnson
©2011

Cover photographs ©iStockphoto.com/David Marchal
©iStockphoto.com/loops7

Cover design by Jessie Nilo Design + Illustration.

No part of this publication may be reproduced, stored, or transmitted without permission or as allowed by law.

Library of Congress Control Number: 2011940789
Library of Congress subject heading:
QH325 Life--Origin

BIASC / BASIC Classification Suggestion:
sh85133362 Teleology

ISBN-13: 978-1-937355-03-6
1.0

Published by Big Mac Publishers, November, 2011
www.bigmacpublishers.com Sylacauga, AL
Printed and bound in the USA

Table of Contents

Acknowledgments

The author wishes to thank all who encouraged the writing of this book. Much gratitude is also extended to astrophysicist Hugh Ross who gave permission to use the probabilities listed in Appendix B.

The author thanks the peer professionals who initially reviewed this book and offered invaluable suggestions for improvements and corrections of errors.

This book is a companion to the "Programming of Life" book that highlights the information and computer aspects of life (released in September, 2010). This book builds on the fine-tuning information from the author's "Probability's Nature and Nature's Probability" books.

Email don@scienceintegrity.net any comments or suggestions for error corrections or improvements.
Errata and additions are available at Website www.scienceintegrity.net.

See www.programmingoflife.com for information on videos based on the Programming of Life books.

Introduction

From childhood, the author was extremely interested in science, devouring books on anything scientific. His love of science led to a Ph.D. in Chemistry from Michigan State University. During his education, views of the American Chemical Society's "From Molecules To Man" program were totally accepted. At that time, he believed anyone not accepting the "proven" chemical and biological evolutionary scenarios was of the same mentality as someone believing in a flat earth. He willingly confronted anyone doubting the evolutionary scenarios, relying on the "facts" presented during his training to promote those scenarios. It is significant that nearly everything in the chemical evolution scenario has been proven false by modern science.

After earning a second Ph.D. in Computer and Information Science from the University of Minnesota, and seriously considering the information in life (bioinformatics), it was realized that the scenarios proposed so far completely fail as being feasible. The "Programming of Life" (PoL) book was written to address the shortcomings of trying to account for life's non-physical information by purely physical interactions of chemicals. It discussed three types of information that are important in life. Shannon information (really probabilistic uncertainty) was shown to be important for code translation (semiotics), and Shannon channel capacity rules out a simpler coding alphabet than that in current life. Functional information was shown to be much more important than mere probabilistic complexity since all aspects of life require functionality. Prescriptive information is an algorithm or recipe to carry out to accomplish the desired task. This is the most important information of life as each living cell has thousands (or millions) of real computer programs executed by thousands of computers. The concept that the cell's software or semiotic coding systems are only metaphors or "computer-like" has been scientifically falsified by the experimental replacement of the DNA software of a bacterium [Gib10]. A cell's hardware and software are experimentally verified realities, as stated by Craig Venter, *"It certainly changed my views of definitions of life and how life works... Life is basically the result of an information process, a software process. Our genetic code is our software"* [Ven10] (note: quotes are italicized, indicating author and year).

While physics and chemistry are physical sciences whose interactions are wholly determined by physicality given any starting constraints, biology is an information science since all of the defining characteristics of biology are controlled by life's information processing systems. Biology isn't just complicated chemistry, since it involves coded messages (semiotics) [Bar03, Bar08] and coded algorithmic prescriptive instructions (instantiated computer programs) [Abe08, 1/1/09, 1/4/09]. These information realities in life are downplayed or ignored by most materialists, since functional information has no feasible cause from physicality (though infeasible scenarios have been speculated). When addressed at all, the informational aspects are usually treated as metaphors or analogies, rather than realities. The PoL book ended by calling for scientific answers to issues raised by life's extreme cybernetic complexity, with millions of interacting co-dependent structures and components. *"Some of the specific problems that require explanation before propagating naturalistic speculations as science include the following.*

How did nature write the prescriptive programs needed to organize life-sustaining metabolism? Programs are shown by computer science to require a formal solution prior to implementation. How did inanimate nature formally solve these complex problems and write the programs? How did nature develop the operating systems and programming languages to implement the algorithms? How did nature develop Turing machines capable of computational halting? How did nature develop the arbitrary protocols for communication and coordination among the thousands (or millions) of computers in each cell?

How did nature develop multiple semiotic coding systems, including the bijective codon-based coding system (for symbolic translation) that involves transcribing, communicating, and translating the symbolic triplet nucleotide block-codes into amino acids of the proteins? How did nature develop alternative generation of such messages using techniques such as overlapping genes, messages within messages, multi-level encryption, and consolidation of dispersed messages? A protein may obtain its consolidated prescriptive construction instructions from multiple genes and/or from the "junk" DNA, sometimes with over a million nucleotides separating the instructions to be combined.

How did nature defy computer science principles by avoiding software engineering's top-down approach required for complex programming systems? How did nature produce complex functional programs without planning by randomly modifying existing algorithms? How did multiple such programs become simultaneously modified to result in the production of irreducibly complex structures?" [Joh10p83-84]

The focus of this book will be the requirements of the physical constants and properties in order for life, with its programming, to exist. As with PoL, this book will deal with science, not philosophy or theology. There are undoubtedly philosophical or theological ramifications raised by the scientific findings, but no stance on those will be taken as that would traverse outside the realm of science. It is important to realize that empirical science does not address issues such as "why" (purpose) or "how" (mechanism), but merely "that." For example, the law of gravity has proven to be extremely accurate for determining intersections of moving objects in 3-dimensional space, such as landing on Mars. Science does not yet know "how" gravity works (gravitons have been proposed, especially as part of string theory), and may never know "why" it works, but science has determined "that" it works (every time).

1 Math Basics: Exponents and Probability

This short chapter may be skimmed by those already familiar with the topics, but to understand what the numbers mean in the rest of this book, it is vital to understand representation and concepts like "possible," "impossible," and "infeasible" presented in this chapter.

Scientific or exponential notation is convenient when expressing very large or very small numbers. The Richter scale for earthquake magnitude is based exponentially, so that a magnitude 5 earthquake is 100 times (2 orders of magnitude) as strong as a magnitude 3 and 100 times weaker than a magnitude 7 quake. Since 6 is 20% higher than 5, if one is thinking linearly, rather than exponentially, one may visualize a magnitude 6 quake as only 20% stronger than a magnitude 5 quake, instead of the 900% stronger (10 times) that it really is. Exponential examples (take note of the exponent) include:

$$4,000,000,000 = 4 \text{ billion} = 4 \times 10^9$$
(count digits to the right of the first digit)

$0.000001 = 1 \text{ millionth} = 1/10^6 = 10^{-6}$ (count right of the "." including the first non-zero digit)

If all you need is a "feel" for the numbers, use the exponent as the number of zeros following or preceding (if negative) the number. Each unit change in exponent is a factor of ten more or less (all of these have the same value): $10^9 = 10 \times 10^8 = 10^{10}/10 = 1000 \times 10^6 = 10/10^{-8}$.

Numbers expressed exponentially may appear to be considerably different than what they represent. For example, a googol (not to be confused with the Google search engine) is 10^{100}, but is physically a totally hypothetical number since the maximum estimate of number of

1

atoms in the Universe is 10^{80} [Sag79] (most estimates are 10^{76} - 10^{78}). There is not a googol of anything physical (except maybe light photons) in the known Universe.

The law of probability expresses the likelihood of a particular outcome from within the set of possible outcomes. Probability has a range of 0 (impossible) to 1 (certain, 100%). Rolling a die has a probability of 1/6 (i.e. 1-in-6) for any particular number. Rolling a five 10 times in a row results in a 1/6 probability of a five on the next die roll since chance has no causative effect. Chance expresses likelihood, not cause – nothing is "caused by chance." The probability of selecting a particular atom of the Universe randomly is 10^{-80}. The probability of rolling ten ones in a row is $(1/6)^{10} = 1.6 \times 10^{-8}$. The probability of tossing 20 consecutive heads is 0.5^{20} or 9.53×10^{-7}.

Note that increasing the number of attempts for something that is impossible will not increase its likelihood since the probability of failure is $(1 - 0)^n = 1$ for any n, so the probability of success is still 0, falsifying statements like biochemist George Wald's declaration, *"Given so much time, the 'impossible' becomes possible... One has only to wait: time itself performs the miracles"* [Wal55]. Likewise, the probability of a certainty doesn't decrease by having more trials since the probability of failure is $(1 - 1)^n = 0$ and $1 - 0$ is still 1 (certain). If the probability isn't 0, theoretically it could happen. For example, the probability of throwing 300 consecutive heads is 0.5^{300} or 4.91×10^{-91}. Although the desired pattern could happen on the first attempt, it is not likely to happen during humanity's existence.

A possible outcome becomes probable when its probability is at least 0.5 since any lower probability makes it more likely not to happen. For example, in rolling a pair of dice, it is probable that the sum will be greater than 6 since there are 21 of the 36 possible combinations that add to greater than 6, so that the probability is $21/36$ (= 0.583) of obtaining a sum higher than 6.

If a die is rolled 100 times with each roll recorded, a very improbable pattern of digits will result, with $P = 1.5 \times 10^{-78}$ (= 6^{-100}) probability for that pattern. Some have argued that just like this pattern happened, life (though improbable) also happened by chance. The fallacy is that if any outcome is acceptable, each roll had a probability of 1 of being correct. As shown, increasing the number of trials for a certainty does not reduce its probability, so $P = 1^{100} = 1$. The number of trials (n) to make a repeat of the pattern probable may be found by waving Taylor's magic wand over the math expression to yield $n = 4.6 \times 10^{79}$. If a roll can be done in one second, the first improbable pattern occurs after only 100 seconds. The repeat of that pattern becomes probable only after 1.4×10^{70} centuries, which is approximately 10^{62} times longer than the oldest estimate of the Universe's age.

Dictionaries (e.g. – Random House©, Inc. 2006 and The American Heritage® Dictionary of the English Language, Fourth Edition) give definitions of infeasible as impracticable or unworkable, *"not capable of being carried out or put into practice."*

In a recent peer-reviewed paper, Abel notes, *"combinatorial imaginings and hypothetical scenarios can be endlessly argued simply on the grounds that they are theoretically possible. But there is a point beyond which arguing the plausibility of an absurdly low probability becomes operationally counterproductive"* [Abe09].

He then calculates criteria and notes that the Universal Plausibility Principle (UPP) states that *"definitive operational falsification"* of any chance hypothesis is provided by inequalities based on the probabilistic resources ranging from 10^{70} chemical reaction on the Earth to 10^{140} quantum interaction in the Universe. If a scenario fails to meet the plausibility inequality standard, *"the hypothetical notion should be declared to be outside the bounds of scientific respectability. It should be flatly rejected as the equivalent of superstition."* For origin of life on Earth, any scenario with a probability less than 10^{-70} is shown to be falsified scientifically (infeasible). Falsification is a criterion that would show a scientific theory is false if the criterion is shown to be true [Pop63].

Science needs a reality-check if origins are to be studied as science. This applies to the origin of mass and energy, the origin of life, and the origin of species. In what other science disciplines would outcomes be published as science if those outcomes had demonstrable probabilities of less than 10^{-100}? When "it's possible that ..." is used, scientists must verify that the pronouncement is indeed possible using known science, as opposed to really meaning "it may be speculated that..." Feasibility also needs to be verified using scientific principles described in the previous paragraph. The feasibility cut-off may vary depending on whether quantum, physical, or chemical interactions are involved, but there is a point where credulity is stretched beyond the breaking point, making science look "foolish" if persisting in treating such paths as pertinent. Without such safeguards, the public will be misled to believe something is science, as opposed to some scientist's speculation or belief.

2 Mass and Energy: Source and Fine-Tuning

The origin of the mass and energy of the Universe has been a topic of much speculation and has continued to be elusive to known science. There are a number of reasons for this, including the non-repeatability of this historic event and the fact that known natural laws fail to account for it. Appendix A gives a brief overview of some of the speculations on origins, pointing out that all such speculation falls outside science as we know it. None of the commonly proposed models can be proved or disproved by known science since known science cannot account for the origin. Each philosophical or theological belief about the origin involves unprovable assumptions that are not verifiable or falsifiable, and would not be bound by known science.

A supernatural belief typically involves powerful being(s) outside the Universe. Quantum fluctuation of "nothing" involves an unseeable eternal "ether" capable of spontaneously generating the Universe. An eternally-existing Universe violates the law of increasing entropy and fails to account for the Universe's increasing rate of expansion. Multi-verse collisions involve innumerable unseeable universes with unseeable dimensions. It is no more scientific to believe in a "scientific-sounding" natural source than to believe in a supernatural source. Each model requires "beliefs," as opposed to empirically determined science. Regardless of how the Universe came into existence, its properties (including its fine-tuned nature) can be empirically examined and evaluated using known science. Although fine-tuning probably wouldn't be surprising for a supernatural source, many of the fine-tuned arguments assume a "Big Bang" scenario for the origin of our Universe, regardless of the ultimate source (supernatural, oscillating, quantum tunneling, etc.).

"Fine-tuning" of the Universe allows life on earth to exist. Incompatibility would result by changing any one of hundreds of parameters. This chapter presents an overview, with specific life chemicals' requirements in chapter 3. Appendix B has an extensive list of 501 parameter requirements (from an analysis of 658 technical papers), each with the probability that a feature will fall into the required range to support simple life [Ros09]. These data indicate that there's less than 1 chance in 10^{311} that even one life-support body would occur anywhere in the Universe for bacteria to exist for as long as 3 months.

The physical constants for weak and strong nuclear forces, electromagnetic and gravitational forces, ratios of forces and electron/proton masses, and properties of neutrons are all extremely critical. For example, if the charge of the proton and electron differed by 1 part per billion, an object the size of a person would fly apart. If the ratio of the electromagnetic to gravitational force were decreased by 1 in 10^{40}, only stars larger than the Sun could occur. The expansion rate, mass, density, and age of the Universe are also critical, as is our position within the solar system, galaxy, and Universe. For example, our solar system is "just right" within the spiral Milky Way with orderly orbits for the right mix of chemicals heavier than helium to form. The Earth's size, orbit, tilt, rotation, magnetic field, atmosphere, and composition are highly unlikely and yet life-critical. For example, a more massive Earth would have more volatile compounds like water, CO_2, and methane, and would be subject to more life-threatening meteor impacts due to higher gravitational attraction. A less massive Earth would quickly lose its atmosphere and probably have an erratic orbit incompatible with life. The Earth's atmosphere allows radiation necessary for life (e.g. for photosynthesis) to enter, but blocks harmful radiation. Small changes in the Earth's orbit or tilt would make the temperature variations inhospitable to life, especially biodiversity.

Astrophysicist Paul Davies states *"There is for me powerful evidence that there is something going on behind it all... It seems as though somebody has fine-tuned nature's numbers to make the Universe"* [Dav88p203]. Stephen Hawking states concerning the constants of physics: *"The remarkable fact is that the values of these numbers seem to have been very finely adjusted to make possible the development of life... It seems clear that there are relatively few ranges of values for the numbers that would allow for development of any form of intelligent life"* [Haw88].

Nobel laureate Steven Weinberg reflects on *"how surprising it is that the laws of nature and the initial conditions of the universe should allow for the existence of beings who could observe it. Life as we know it would be impossible if any one of several physical quantities had slightly different values"* [WeiSA]. Physicist David Deutsch notes, *"If we nudge one of these constants just a few percent in one direction, stars burn out within a million years of their formation, and there is no time for evolution. If we nudge it a few percent in the other direction, then no elements heavier than helium form. No carbon, no life. Not even any chemistry. No complexity at all"* [Deu06]. If the strong nuclear force *"were just 2%*

weaker or 0.3% stronger than it actually is, life would be impossible at any time and any place within the universe" [Swi90]. Physicist John Barrow writes, *"small changes in the electric charge of the electron would block any kind of chemistry"* [Bar80]. Theoretical physicist Lee Smolin has calculated the probability of star (including our Sun) formation from random parameters as 10^{-229} [Smo97]. While admitting *"life as we know it on Earth would not exist if several of the parameters of physics were different from their existing values"* [Ste07p146], physicist Victor Stenger speculates that there is a *"possibility that an appreciable number of planets exist with conditions that, while unsuitable for our form of life, can support some kind of life"* [Ste07p144]. No scientific proof is offered that life as we don't know it is "possible" (i.e.– non-zero probability), so such statements amount to wishful thinking and pure speculation, not part of science.

"The small value of the cosmological constant is telling us that a remarkably precise and totally unexpected relation exists among all the parameters of the Standard Model of particle physics, the bare cosmological constant and unknown physics" [Abb91]. This constant needs a precision of one part in 10^{120} [Mic99]. To get a feel for what that precision means, if one could measure the weight of a 220 pound man to the correct number of grams, a precision of one part in 10^5 would be required. To measure the mass of the Universe with an accuracy down to the mass of an electron, a precision of one part in 10^{74} would be required. A precision of one part in 10^{120} requires 10^{46} times more precision than that!

Nima Arkani-Hamed has attempted *"to explain why things that appear to be finely, even heroically, tuned*

actually are not. One possibility, he said, is that our universe is not unique but is only part of a vast 'landscape' of universes. If there are huge numbers of universes, perhaps 10 to the 500th power by one estimate, then it is no great stretch to imagine that at least one of them —ours— wound up having extremely small amounts of observed vacuum energy and a weak force that operates on a scale much smaller than expected." [AAAS05].
Note that "possibility" is used without non-zero probability proof for this non-scientific speculation. In analyzing the precision of the original Big Bang entropy, Penrose calculates *"An accuracy of one part in $_{10}10^{123}$...the precision needed to set the universe on its course"* [Pen89]. Note: 10^{123} is a 1 followed by 123 zeros, $_{10}10^{123}$ is a 1 followed by 10^{123} zeros! That number is so far beyond anything real (remember there are less than 10^{80} atoms in the Universe) that is difficult to imagine.

Hawking and Mlodinow describe how miraculous it is that the laws of physics allow for a Universe that is hospitable for life, in which the Universe has an excess of matter over antimatter and galaxies with stars (with planets) that last billions of years. They attempt to "explain" the observation using "M-theory," which unifies gravity with the other weak and strong nuclear and electromagnetic fundamental forces. They predict seven additional space dimensions [Haw10]. *"Besides the absence of any compelling experimental evidence for M-theory, there is another difficulty — its predictions are far from unique. There are 10^{500} different ways to curl up the extra seven dimensions and hide them, and how they curl up determines the fundamental constants and what we four dimensional creatures see as the laws of physics.*

So even if M-theory is the only theory of everything available, there remain 10^{500} possibilities for the laws of physics we observe. Thus, say Hawking and Mlodinow, there is no miracle — inflation plus M-theory equals multiverse. Our special Universe is a selection effect: all possibilities have been tried and we find ourselves in the only kind of inflationary patch that can support our existence... Hawking and Mlodinow argue that negative gravitational potential energies allow something to arise from nothing — but that still begs the question of why there is space, time and M-theory at all" [Tur10]. It also doesn't explain the origin of gravity, on which their unscientific speculation depends. Furthermore, it doesn't explain the fine-tuning in **OUR** Universe, which is the only one that we can observe.

Perhaps someday science will come up with a verifiable undirected natural model to explain the apparent fine-tuning that is evident, and science should certainly seek such knowledge. At this point it would be incorrect to state that "it's possible that science will come up with an undirected natural solution to this question" since that would assume that non-zero probability of such a solution were proven. Hoyle was an atheist when he wrote *"A common sense interpretation of the facts suggests that a superintellect has monkeyed with the physics ... The numbers one calculates from the facts seem to me so overwhelming as to put this conclusion almost beyond question"* [Hoy81U]. Davies states *"If the world's finest minds can unravel only with difficulty the deeper workings of nature, how could it be supposed that those workings are merely a mindless accident, a product of blind chance?"* [Dav84]

3 Chemicals of Life

While life uses the laws of chemistry and physics, those laws cannot define or explain life any more than the rules of grammar that were used during the preparation of this book define its content. This chapter will review the major components of life, and then analyze the requirements of the chemicals involved.

Amino acids are the building-blocks of life. Each is an organic molecule that has a carboxylic acid ($-CO_2^-$) and an amine ($-NH_3^+$) group attached to the same carbon atom. The second carbon has another chemical side-group (or side-chain) attached. There are 20 main amino acids for life.

A peptide bond results during a catalyzed dehydration reaction in which the carboxylic group of one acid reacts with the amino group of another. Multiple such reactions link amino acids together into polypeptide chains. Functional polypeptide chains longer than approximately 100 amino acids are referred to as proteins which are synthesized in and used by living cells. Enzymes are catalytic proteins that have special slots to hold other molecules to make the chemical reactions of life feasible. The longest known biochemical *"half-time - the time it takes for half the substance to be consumed - is 1 trillion years, 100 times longer than the lifetime of the universe. Enzymes can make this reaction happen in 10 milliseconds...Without catalysts, there would be no life at all... It makes you wonder how natural selection operated in such a way as to produce a protein that got off the ground as a primitive catalyst for such an extraordinarily slow reaction"* [Wol03].

11

Deoxyribonucleic acid (DNA) contains the genetic information of a cell, including the information for constructing proteins (including enzymes) and RNA (Ribonucleic acid). Information is organized as genes (often one gene per protein). Chemically, DNA is a long polymer of nucleotides, each nucleotide being a deoxyribose sugar molecule, one phosphate group, and one base. Nucleotides are joined by ester bonds so that the sugars and phosphates form the DNA backbone, with the bases sticking out to form hydrogen bonds with a second DNA strand to form a double helix [Wat53]. The four bases used in DNA are adenine (A), cytosine (C), guanine (G) and thymine (T), with base pairing between the helix chains always being G-C or A-T. Uracil (U) usually takes the place of T in RNA. The sequence of the four bases along

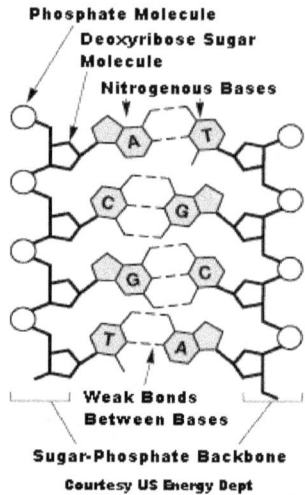

Phosphate Molecule
Deoxyribose Sugar Molecule
Nitrogenous Bases
Weak Bonds Between Bases
Sugar-Phosphate Backbone
Courtesy US Energy Dept

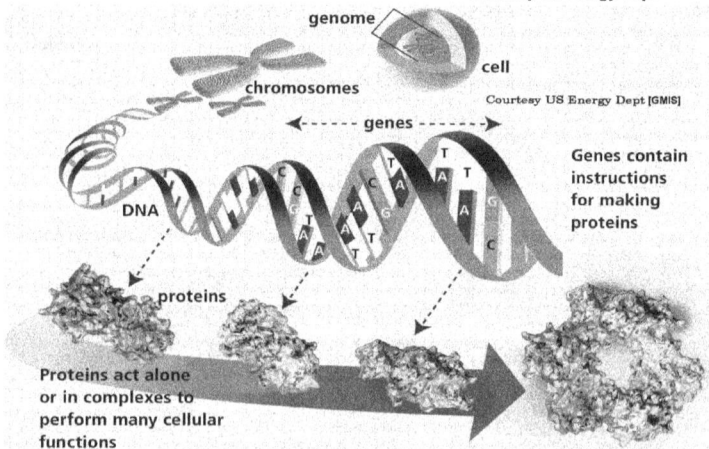

genome
cell
chromosomes
Courtesy US Energy Dept [GMIS]
genes
DNA
Genes contain instructions for making proteins
proteins
Proteins act alone or in complexes to perform many cellular functions

the backbone encodes the genetic prescriptive instructions used in the development and functioning of all known living organisms.

The four main elements of life are: Hydrogen (H, 59%), Oxygen (O, 24%), Carbon (C, 11%), and Nitrogen (N, 4%). The other elements making up about 2% of the cells' content include:

Bromine – Br, for pigmentation and defense

Calcium – Ca, in bone, muscle, messaging

Chlorine – Cl, for photosynthesis and digestion

Chromium – Cr, for metabolism.

Copper – Cu, for oxygen-carrying pigment

Fluorine – F, for tooth enamel development

Iodine – I, in thyroxine hormone

Iron – Fe, for oxygen carrying (in hemoglobin)

Magnesium – Mg, in chlorophyll and enzymes

Manganese – Mn, needed for some enzymes

Phosphorus – P, in DNA/RNA and ATP

Potassium – K, for nerve impulse generation

Selenium – Se, in many enzymes

Silicon – Si, in bacteria , grass, leaves

Sodium – Na, for nerve conduction

Sulfur – S, in most proteins

Zinc – Zn, in alcohol oxidizing enzyme

Carbon has many qualities that make it uniquely suitable as a major component in life. It forms bonds with many other elements to produce the chemicals of life, including those of the information-holding structures. These bonds are stable enough to withstand harmful chemical and physical assaults, yet not so strong so as to prevent many different kinds of reactions, especially with enzymatic assistance. Carbon compounds are found in all three states: gas (e.g. CO_2 and CH_4), liquid (e.g. body fluids), and solid (e.g. protein structures and DNA).

Hydrogen is the major component of life. It would not exist if the strong nuclear force constant were larger, and if that constant were smaller, hydrogen would be the only element in the Universe. Hydrogen not only forms bonds with carbon to form hydrocarbons, but combines with oxygen to form the compound most critical to support life, water, which is used as a universal solvent to carry nearly all other compounds needed by life. *"Water's life-giving properties exist on a knife-edge. It turns out that life as we know it relies on a fortuitous, but incredibly delicate, balance of quantum forces. Water is one of the planet's weirdest liquids, and many of its most bizarre features make it life-giving"* [Gr011].

Water has many unique properties critical for life support. For example, ice is less dense than liquid water, so that bodies of water freeze from the top, rather than becoming a solid block from the bottom up. Ice insulates the liquid below. Water's heat of vaporization regulates temperatures of bodies of water to support life therein, and also of living bodies (e.g. via sweat). Water's high surface tension permits capillary action to transport liquid to the tops of plants. Water also plays important roles in the formation of geological structures (e.g. glacial action and freeze/thaw cycles to break down rocks), and climate control. Water is involved or produced by many reactions of life. For example, a molecule of water is removed when two amino acids are joined during protein (or other polypeptide) construction. Since liquid water is essential for life, there are very few zones of habitability (based on host star's characteristics, orbit, atmosphere, etc.) possible that wouldn't result in a run-away greenhouse effect or complete freezing. In fact, liquid water is an anomaly, since the boiling-point trends of other hydrogen-containing compounds would predict H_2O "should" boil at -100° C, not the observed +100° C (this is due to strong hydrogen bonding between molecules because of bond angle).

Oxygen is critical for metabolism in most life. Multicellular life typically transports oxygen via hemoglobin in blood, using an easily-reversible complex that requires iron. If the heme-oxygen bonds were stronger, cells would not be able to access the oxygen, and if it were weaker, the lungs wouldn't be able to oxygenate the blood. Note that carbon monoxide poisoning occurs because CO forms a strong heme bond, leaving insufficient hemoglobin to carry oxygen.

Nitrogen is essential for all forms of life, being a major component of proteins and genes. Although nitrogen makes up about 80% of the atmosphere of earth in the form of molecular N_2, that form cannot easily be absorbed by living cells. Only some nitrogen fixing bacteria (e.g. blue-green algae) can split these two atoms apart to form ammonium salts and nitrates. All higher plants and animals rely upon these nitrogen fixers to give them nitrogen in a form they can use, often by ingesting food that traces back to the fixers.

Wavelengths of radiation important for life's chemicals are also finely-tuned at both the source (Sun) and destination (Earth's chemicals). If the energy of photons were different, either molecules could be torn apart by radiation, or photosynthesis couldn't occur. The portion of the electromagnet spectrum vital to life is only 10^{-25} of that available, and the Sun's emission (determined by its temperature and composition) is ideal for enabling chemical reactions (photon absorption raises an electron's energy level to a specific required level).

Heavier elements can form by nuclear fusion (near a galaxy's center) in which a new stable nucleus combines two other nuclei. Two helium-4 nuclei can form a beryllium-8 nucleus. A ^{4}He and ^{8}Be can form ^{12}C, and a ^{12}C and a ^{4}He can form a ^{16}O. The half-life of ^{8}Be is a critical 10^{-15} seconds, which makes ^{12}C formation "almost" a three-body collision. If ^{8}Be were less stable, nothing higher (e.g. ^{12}C) would form, and if it were more stable, production of higher chemicals would proceed so rapidly that no life-required chemicals would remain. If nuclear energy ratio of ^{12}C to ^{16}O were larger, there would be insufficient oxygen, and if smaller, insufficient carbon.

As indicated in Appendix B, the formation of the chemicals needed for life, as well as the physical constants for science are extremely unlikely. The materialistic scenarios proposed thus far are in fact, operationally falsified [Abe09]. This leaves science in a quandary when trying to explain the apparent fine-tuning that is scientifically verifiable in nature.

4 Going Where Data Lead

Biological Philosopher Ronald Brady wrote, *"By making our explanation into the definition of the condition to be explained, we express not scientific hypothesis but belief. We are so convinced that our explanation is true that we no longer see any need to distinguish it from the situation we were trying to explain. Dogmatic endeavors of this kind must eventually leave the realm of science"* [Bra85].

"Science" has its root in the Latin word "Scientia," meaning knowledge or truth. *"Science... accumulated and accepted knowledge that has been systematized and formulated with reference to the discovery of general truths or the operation of general laws: knowledge classified and made available in work, life, or the search for truth: comprehensive, profound, or philosophical knowledge; especially knowledge obtained and tested through the use of the scientific method"* [Web93]. Nobel prize winner Linus Pauling said: *"Science is the search for truth, the effort to understand the world; it involves the rejection of bias, of dogma, of revelation, but not the rejection of morality... One way in which scientists work is by observing the world, making note of phenomena, and analyzing them"* [PauWeb]. It would be nice to believe that presuppositions and biases are not involved with science, and that scientists would follow the evidence, wherever it took them. It is healthy and proper to thoroughly examine claims that purport to be scientific because anyone can speculate anything, but that doesn't make it true. What should not be tolerated is refusal to evaluate the scientific merits because of philosophical or theological views. For example, biologist Richard Lewontin writes *"Our willingness to accept scientific claims that are against common sense is the key to an understanding of the real struggle between science and the supernatural. We take the side of science in spite of the patent absurdity of some of its constructs, in spite of its failure to fulfill many of its extravagant promises of health and life, in spite of the tolerance of the scientific community for unsubstantiated just-so stories, because we have a prior commitment, a commitment to materialism. It is not that the methods and institutions of science somehow compel us to accept a material explanation of the phenomenal world, but, on the contrary, that we are forced by our a priori adherence to material causes to create an apparatus of investigation and a set of concepts that produce material explanations, no matter how counter-intuitive, no matter how mystifying to the uniniti-ated. Moreover, that materialism is absolute, for we cannot allow a Divine Foot in the door"* [Lew97].

Most scientists apparently hold materialism as the only basis for science. By contrast, when the sufficiency of naturalism was questioned in a presentation at a Bioinformatics conference [Joh04], the reception was friendlier than anticipated, with dozens of scientists expressing support or non-negative interest (less than a third were negative) in the concepts. It seems that scientists who have been actually investigating the vast information of life are more willing to consider models that many mainstream scientists find objectionable. This book is about science, and it is hoped that it will be evaluated based on science, and not philosophical beliefs. This chapter is not meant to advocate any particular belief, but to point out beliefs that prevent many scientists from engaging in an honest evaluation of the scientific content.

Although many aspects of fine-tuning are compatible with various theological views, and there are undoubtedly philosophical and theological ramifications, those fall outside the realm of empirical science, so this book will make no pronouncement in support of any particular view. The US Supreme Court has held that the implications of material alone do not make a religion even though those implications *"coincide or harmonize with the tenets of some or all religions"* [Sup80]. It also ruled *"The Establishment Clause stands at least for the proposition that when government activities touch on the religious sphere, they must be secular in purpose, evenhanded in operation, and neutral in primary impact"* [Sup71]. A recent US Appeals Court ruling rejected the claim that *"Texas Education Agency's ('TEA') neutrality policy constitutes an establishment of religion, in violation of the First Amendment's Establishment Clause. Because we find no evidence to support the conclusion that the principal or primary effect of TEA's policy is one that either advances or inhibits religion, we conclude that the policy does not violate the Establishment Clause. As such, we affirm the decision of the district court"* [App10]. The National Center for Science Education supported the suit, claiming that the TEA policy was endorsing "Creationism" (which it wasn't, but any evidence potentially bringing Darwinism into question is thought by many naturalists to be a religious stance). Excellent legal reviews of court rulings are available for further consideration [Lus09, Cal09].

Some may object to the concept of fine-tuning

because of religious compatibility, which should mean they should be equally adamant against Darwinism and naturalism, since they are compatible with Atheism, Secular Humanism, and Naturalism, which are definitely religions, as affirmed by the Supreme Court: *"Among religions in this country which do not teach what would generally be considered a belief in the existence of God are Buddhism, Taoism, Ethical Culture, Secular Humanism, and others"* [Sup61]. It also indicated that *"religious beliefs... are based... upon a faith, to which all else is subordinate or upon which all else is ultimately dependent.... Some believe in a purely personal God, ... others think of religion as a way of life"* [Sup65]. This Supreme Court criterion makes "physicality is all there is" a religious belief since it cannot be proven and all else is subordinate to it. A US Appellate court also affirmed that *"Atheism is religion, and the group... was religious in nature even though it expressly rejects a belief in a supreme being"* [App05]. Because of these court rulings, one must use care when presenting any speculative purely physical naturalistic scenario, since transgression could be an establishment violation (with the same warnings as teaching Creationism would bring). Presenting only verifiable science, avoiding unverified speculations, should be safe. Atheist and evolutionary biologist David Sloan Wilson admits, *"many scientific theories of the past become weirdly implausible... [and] are a greater cause for concern because they do a better job of masquerading as factual reality. Call them stealth religions"* [Wil07].

A supernatural source is erroneously believed by many to be required for acknowledging fine-tuning. It is worthwhile to look at some pertinent statements, such as by evolutionist Michael Denton. *"It is important to emphasize at the outset that the argument presented here is entirely consistent with the basic naturalistic assumption of modern science – that the cosmos is a seamless unity which can be comprehended ultimately in its entirety by human reason and in which all phenomena, including life and evolution and the origin of man, are ultimately explicable in terms of natural processes"* [Den98pxviii]. And later in that book: *"Evolution was accepted in the nineteenth century not because it explained everything perfectly but because it accounted for the facts better than any other theory...The idea that the cosmos is a unique whole with life and mankind as its end and purpose makes sense and illuminates all our current scientific knowledge. It makes sense of the intricate synthesis of carbon in the stars, of the constants of physics, of the properties of water, of the cosmic abundance of the elements... No other world view comes close. No other explanation makes as much sense of all the facts"* [Den98p385].

Fred Hoyle was an atheist when he wrote *"The enormous information content of even the simplest living systems... cannot in our view be generated by what are often called 'natural' processes... There is no way in which we can expect to avoid the need for information, no way in which we can simply get by with a bigger and better organic soup, as we ourselves hoped might be possible... The correct position we think is... an intelligence, which designed the biochemicals and gave rise to the origin of carbonaceous life... This is tantamount to arguing that carbonaceous life was invented by noncarbonaceous intelligence"* [Hoy81E].

21

The scientific "establishment" has inhibited opposing scientific views, denying funding, publishing, or jobs to those not toeing the mainstream dogma [Ber08, Exp08, Cro10]. Dissenters to big bang cosmology note that opposing view *"development has been severely hampered by a complete lack of funding. Indeed, such questions and alternatives cannot even now be freely discussed and examined... all the peer-review committees that control them are dominated by supporters of the big bang. As a result, the dominance of the big bang within the field has become self-sustaining, irrespective of the scientific validity of the theory"* [Ope04]. *"Science is a process within a community of people looking for truth; string theory is just a faith-based science... for twenty years it has soaked up most of the funding, attracting some of the best scientific minds, and penalizing young physicists for pursuing other avenues"* [Smo07].

To expound some scientifically mainstream views, Pagels writes that so powerful is the scientific-experimental *"method that virtually everything scientists know about the natural world comes from it. What they find is that the architecture of the universe is indeed built according to invisible universal rules, what I call the cosmic code-the-building code of the Demiurge. ... It must be the work of an Alien Intelligence! ... Whether God is the message, wrote the message, or whether it wrote itself is unimportant in our lives. We can safely drop the traditional idea of the Demiurge, for there is no scientific evidence for a Creator of the natural world, no evidence for a will or purpose that goes beyond the known laws of nature. Even the evidence of life on earth, which promoted the compelling 'argument from design' for a*

Creator, can be accounted for by evolution... So we have a message without a sender" [Pag88]. Dawkins writes, *"The feature of living matter that most demands explanation is that it is almost unimaginably complicated in directions that convey a powerful illusion of deliberate design"* [Daw01], and *"The kind of explanation we come up with must not contradict the laws of physics. Indeed it will make use of the laws of physics, and nothing more than the laws of physics"* [Daw96Bp15].

To expound some revolutionary views, Paul Davis (professor of theoretical physics) writes: *"The really amazing thing is not that life on Earth is balanced on a knife-edge, but that the entire universe is balanced on a knife-edge, and would be total chaos if any of the natural 'constants' were off even slightly... even if you dismiss man as a chance happening, the fact remains that the universe seems unreasonably suited to the existence of life -- almost contrived -- you might say a 'put-up job'"* [DavWiki].

"Biologists must be encouraged to think about the weaknesses of the interpretations and extrapolations that theoreticians put forward or lay down as established truths. The deceit is sometimes unconscious, but not always, since some people, owing to their sectarianism, purposely overlook reality and refuse to acknowledge the inadequacies and the falsity of their beliefs" [Gra77p8]. Hoyle wrote, *"It is ironic that the scientific facts throw Darwin out, but leave William Paley, a figure of fun to the scientific world for more than a century, still in the tournament with a chance of being the ultimate winner... Indeed, such a theory is so obvious that one wonders why it is not widely accepted as being self-evident. The*

reasons are psychological rather than scientific" [Hoy81Ep130].

"A Chinese paleontologist lectures around the world saying that recent fossil finds in his country are inconsistent with the Darwinian theory of evolution... When this conclusion upsets American scientists, he wryly comments: 'In China we can criticize Darwin but not the government. In America you can criticize the government but not Darwin.'... one reason the science educators panic at the first sign of public rebellion is that they fear exposure of the implicit religious content in what they are teaching" [Joh99]. Presidential Medal of Science winner Lynn Margulis notes, *"as far as 'survival of the fittest' goes, ...* [natural selection is] *neither the source of heritable novelty nor the entire evolutionary process...* [making] *Darwinism 'dead,' since there's no adequate evidence in the literature that random mutations result in new species... Natural selection is the failure to reach the potential, the maximum number of offspring that, in principle, can be produced by members of the specific species in question"* [Maz10p257&267]. Michael Polanyi turned to philosophy at the height of his scientific career when he saw how ideologies were being employed to hinder free scientific expression and inquiry. Polanyi argued life is not reducible to physical and chemical principle, but rather, *"the information content of a biological whole exceeds that of the sum of its parts"* [PolWeb]. The abiogenesis argument *"simply says it happened. As such, it is nothing more than blind belief. Science must provide rational theoretical mechanism, empirical support, prediction fulfillment, or some combination of these three. If none of these three are available, science should*

reconsider that molecular evolution of genetic cybernetics is a proven fact and press forward with new research approaches which are not obvious at this time" [Tre04].

Intelligent people interpret data differently, largely based on presuppositions and training. Currently, those espousing the undirected natural processes scenarios are in control of the vast majority of scientific clout. Those holding any other view are largely shunned as being "unscientific." Studies have repeatedly shown that the public holds views that are not compatible with undirected naturalism, yet it is taught as "truth" in public school systems, despite its lack of scientific evidence. Since Atheism, Secular Humanism, and pure naturalism (to which all is subservient) have been confirmed as religions by US courts, and Evolution has been declared a religion by evolutionists, this teaching may be against the First Amendment religious establishment clause. Those who want to promulgate undirected naturalism as unassailable dogma certainly have the right to establish private schools (not tax supported) to carry out that objective. Students currently are being taught half-truths and even falsehoods as if they were true.

There are three topics that should not be included in science unless/until sufficient facts from known science demonstrate feasible scenarios: the origin of the mass and energy of the Universe, the origin of life, and the origin of species. To pretend that a "scientific-sounding" scenario is actually science in these areas does a disservice to both science and the public, and diminishes the reputation of all science by anyone examining the evidence. It would be appropriate to teach why those areas are not science, and that science has no scientific answers to questions of origins. When evolution is taught, its scope should be limited to that which is known, eliminating unwarranted speculation. Evolution has been a knowledge-stopper instead of an enlightener. For example, how long did it take to actually study the vestigial organs to discover that all 180 that were considered useless remnants of evolution are actually useful or even vital? More recently, what for over 20 years was considered "junk DNA" caused by mutations which haven't yet mutated into useful genes, has found many useful purposes, and seems likely to yield more positive results. When alternatives to naturalism are taught, only the empirical scientific evidence (such as fine-tuning, irreducible complexity or functional and prescriptive information) should be taught, with the fact that science cannot know who or what caused the features (obviously any theological implications should not be addressed in a public school).

When reading or hearing findings that purport to be science, be critical. This applies to peer reviewers, students, and the general public (e.g. TV's "science programs"). If "possible" is used, analyze whether or not non-zero probability has been proven using known science. If it hasn't, don't treat the scenario as science, but rather as faith or speculation. A good benchmark for taking seriously any abiogenesis scenario would be if that scenario were awarded the $1 million Origin of Life Prize [OOLprize]. If that prize is awarded, the winning scenario may be considered science. Until then, view any abiogenesis scenario with extreme skepticism. No scenario should be published as science if it fails the feasibility test described on page 3 [Abe09].

6 Epilog

Dawkins writes: *"The more statistically improbable a thing is, the less can we believe that it just happened by blind chance. Superficially the obvious alternative to chance is an intelligent Designer. But Charles Darwin showed how it is possible for blind physical forces to mimic the effects of conscious design, and, by operating as a cumulative filter of chance variations, to lead eventual[ly] to organized and adaptive complexity, to mosquitoes and mammoths, to humans and therefore, indirectly, to books and computers"* [Daw82]. In other words, those espousing undirected natural causes for everything, believe that the original chance formation of the Universe led to chance formation of life which led to humans which led to computers, etc. Tracing the propositional logic backwards leads to chance causing the computer that is being used to record the thoughts (also caused ultimately by chance) for this book. Ignoring the fact that chance has no "causative effect," where is the proof of such a belief? What is the scientific rationale for the belief *"biology is the study of complicated things that give the appearance of having been designed for a purpose"* [Daw96Bp1]? What scientific rationale causes the belief that the Universe's constants that *"appear to be finely, even heroically, tuned"* [AAAS05] is just an illusion?

The limits of science may be philosophical, rather than scientific. *"What are the limits for what can claim to be science? One proposal is methodological naturalism (MN), which requires that scientific theories can postulate only natural causes. What are the limits for* *what MN-Science can claim to explain? If we decide to accept methodological naturalism, a second limit is logically necessary: If an event really does involve a non-natural cause, any explanation of the event by MN-Science (in terms of only natural causes) will be incomplete or incorrect"* [RusWeb]. *"Limiting science to a predetermined set of acceptable explanations naturally begs the question, 'What if there is no natural explanation?' ... Science would forever miss it and would continue to squander intellectual and financial capital on finding naturalistic answers that do not exist. Scientific progress depends heavily upon discovering blind alleys and rejecting failed theories. This is simply the way that science works"* [Bow07]. Naturalism has tried to squeeze the data into prevailing suppositions, refusing to even consider alternate scenarios that are at least as scientific, and refusing to fund or publish those alternatives.

Laws of nature are not just descriptive, but are prescriptive. While a law may only be approximated by empirical human understanding of it, there is an underpinning formal law that controls physicality through its prescriptive information. We can simulate a law's effects using a computer algorithm to characterize the observations, even if we don't understand the underlying mechanism. For example, the attractive force between two masses separated by distance, r, is empirically $f = Gm_1m_2/r^2$, where G is the gravitational constant. Although gravitons have been proposed as a mechanism, they have no scientific verification. Life is filled with numerous actual programs that prescribe life's processes. Even the laws of nature, to which life also conforms, are examples of formal prescriptive information. The scenarios to explain nature's fine-tuning acknowledge that the laws of physical interactions and the constants could have been otherwise (formally contingent). Since there is no scientifically verifiable cause of that formalism, we are left accepting scientific laws as axioms.

Occam's razor (a philosophical principle) *"states that the explanation of any phenomenon should make as few assumptions as possible, eliminating those that make no difference in the observable predictions of the explanatory hypothesis or theory... entities must not be multiplied beyond necessity... 'All other things being equal, the simplest solution is the best.' In other words, when multiple competing theories are equal in other respects, the principle recommends selecting the theory that introduces the fewest assumptions and postulates the fewest entities"* [OccWik].

To recap the arguments, we start with the fine-tuning of the Universe, since origin of mass and energy is scientifically unknowable. Consider just four numbers (ignoring the precision of original entropy of $_{10}\text{-}10^{123}$ [Pen89]): star formation probability of 10^{-229} [Smo97], cosmological constant precision of 10^{-120} [Mic99], the necessary 10^{500} "universes" to make feasible the evident fine-tuning [AAAS05], and the 10^{-311} probability of a single planet capable of sustaining simple life [Ros09]. Each probability falsifies every scenario proposed thus far [Abe09].

As indicated previously [Joh10], for abiogenesis, the probability of forming the simplest form of living organism known is $10^{-340,000,000}$ [Mor79], which is operationally falsified, but there really is no need to do so, since the probability of an undirected source of information contained in life is 0 (impossible) based on life's choice-contingency instantiated algorithms, functional information (including semiotic systems), and alphabet requirements for information transfer. It is infeasible to put a reliable estimate on the probability of undirected production of the tree of life by undirected causes (considering no new net information results by mutations, irreducible complexity, Cambrian explosion, etc.). Since the functional information of human DNA is over 10^7 bits more than the simplest organism, somehow over 10^7 bits would have to be injected into the genome, but there's no known mechanism for that to happen. Using functional information, the ratio of formation probabilities of human to simplest organism has been shown to be less than $10^{-3000000}$.

These findings indicate that science is far from answering the questions of origin of mass/energy, life, or species. Any speculative scenarios are based on unsubstantiated presuppositions, and are really matters of faith, and not science. It should be noted that all science is tentative, so new findings may modify that conclusion which is based on today's known science. At this point, one cannot say "it's possible new findings may produce scientific answers to origins," since that would assume such findings are possible (non-zero probability), which has not been proven by known science. Those disagreeing with this conclusion are invited to show from known science that it is incorrect.

Appendix A: Mass/Energy Origin Scenarios

The law of conservation of mass and energy says that matter and energy cannot be created nor destroyed, but may be converted from one form to another ($\Delta E + c^2\Delta M = 0$, where c is the speed of light). Before Einstein, each change (ΔE or ΔM) was assumed to be 0, but science now knows that mass can be converted to energy, and vise versa. If a gram of mass is converted (ΔM is negative) into energy, such as in an atomic bomb, 9×10^{13} joules of energy are produced. Note that the atomic bomb that leveled Hiroshima was over 160 times less than this with a 6.4 mg loss of nuclear mass. In a nuclear power plant, 2700 MWHrs of electric power is produced per gram of mass converted. If one could convert the mass of a can of soda into energy, it would produce enough electric power for a million homes in the United States for a year. When one considers the 3×10^{69} joules of energy required to produce the estimated 3×10^{55} g mass of the Universe [NASA], one is forced to look beyond the random energy fluctuations that were once proposed as a source. *"The principle of the conservation of energy is considered to be the single most important and fundamental 'law of nature' known to science, and is one of the most firmly established. Endless studies and experiments have confirmed its validity over and over again under a multitude of different conditions"* [You85].

Since undirected natural processes producing "matter and energy from nothing" has been dismissed by virtually all scientists (some do believe there is no such thing as "nothing," however), "eternal existence" of matter and energy has been proposed. This model proposes an oscillating Universe that has always existed, alternating between a Big Bang and a Big Crunch. Although this model complies with the conservation law, there are two major problems. One problem is that the known mass of the Universe is insufficient to cause a collapse since the Universe is expanding [Hub29] at a rate faster than the escape velocity (gravitational forces cannot slow the expansion enough to cause collapse). Some have speculated the existence of "dark matter," which can't be seen, but whose mass would make a collapse possible, that is the Universe's density exceeds the Friedmann critical value. Dark matter estimates have shown densities of less than half this required Friedmann critical value. Recent findings indicate that even the rate of expansion is increasing [Pee03], which means the Universe cannot collapse. "Dark energy" is the proposed "explanation" of this increase in expansion rate. What *is driving this apparently anti-gravitational behavior on the part of the Universe, nobody claims to understand why it is happening, or its implications"* [Ove08]. Three physicists shared the 2011 Nobel prize for verifying the Universe's accelerating expansion.

The Second Law of thermodynamics says that in a closed system, entropy always increases. This is the second difficult scientific problem to accommodate for a Universe of infinite age. Entropy is a measure of randomness (disorganization) or inability to do work (limiting energy transfer between entities). Isaac Asimov has said *"As far as we know, all changes are in the direction of increasing entropy, of increasing disorder, of increasing randomness, of running down. Yet the universe was once in a position from which it could run down for trillions of* *years. How did it get into that position?"* [Asi73] An infinitely old Universe that is not at maximum entropy violates the Second Law because energy is still being transferred within the Universe, which is a closed system by its "uni-" definition. Some have argued that the Universe isn't closed, since it is expanding without limits. "Closed" has nothing to do with physical size, but rather with matter/energy content. In addition, it has been proven that in an open system *"entropy cannot decrease faster than it is exported through the* [open] *boundary, because the boundary integral there represents the rate that entropy is exported across the boundary"* [Sew05]. Note that directed energy can cause a local entropy decrease without violating the Second Law, since the system encompassing that local system would have an entropy increase. An infinitely old Universe would be energy-dead, with no capacity for work, since one result of the Second Law is that perpetual-motion machines are impossible (zero probability). There would be no available energy when randomness is maximum (after infinite time).

Einstein's general theory of relativity (GTR) has become the basis for several proposed speculations. The main set (for each tensor u and v) of equations relating the curvature tensor (R_{uv}), the spacetime tensor (T_{uv}) is given by the simplified partial differential equation $R_{uv} - (1/2) \, g_{uv} \, R = (8\pi G/c^4) \, T_{uv}$, ($g_{uv}$ includes first and second derivatives), where c is the speed of light and G is the gravitational constant. Spacetime is an invisible flowing stream that bends in response to objects in its path, as it carries everything in the Universe along its twists and turns. Smaller objects travel through space that is warped by the larger object. Since a second-order differential equation isn't limited to a particular solution, there are, with a variety of assumed starting points, *"an infinite number of solutions. The solution that is used within the scientific community is that given by Einstein. He chose the mathematically simplest equation that relates matter and energy to the curvature of space-time. But in truth, there is no reason why reality should conform to our desire for mathematical simplicity"* [Ast05]. Some of the solutions have resulted in speculation about many esoteric concepts [Mor88].

GTR is widely accepted because it predicts many experimentally verified results such as black holes [Kor95], gravitational bending of light [Edd20], and gravitational time dilation (slows with increasing field strength) [Pou65]. A black hole has that name since even light cannot escape its strong gravitational field as it sucks into itself the surrounding spacetime. Some GTR-allowed possibilities are hypothetical. A wormhole is a hypothetical shortcut through spacetime, allowing apparent faster-than-light travel. A white hole is a hypothetical time reversal of a black hole, ejecting matter from its event horizons. A singularity is a hypothetical place with infinite gravitational field where the curvature of space-time and the density of matter become infinite.

That GTR seems to produce verifiable results with some starting assumptions does not mean that all possible starting points are indeed "possible" (with non-zero probability in reality). For example, Stenger treats negative time as a GTR-allowed possibility – *"the equations of cosmology that describe the early universe apply equally for the other side of the time axis"* [Ste07p126] -- despite almost universal acceptance that time before Planck time (5.4×10^{-44} sec, the theoretically smallest measurable time unit in quantum mechanics) is unknowable. He and others assume that *"there is no such thing as nothing"* since GTR allows for "empty" space to have "vacuum energy," also known as "ether," which could spontaneously produce (without cause, via quantum tunneling) the mass/energy and the negative gravitational energy of the Universe via spontaneous symmetry breaking. In quantum theory, quanta can exhibit both particle and wave properties. When confronting an energy barrier too high to ascend normally, a quantum can be described by a wave-function that has probabilities of positioning it on either side of the barrier. Quantum tunneling results if the function describing the particle indicates it is on the other side of the barrier.

Perhaps a reality check is in order. Eternal existence of such space would require that it would be at maximum entropy. Speculation [Ste07] that this maximum entropy system gave rise to the physical Universe that is not at maximum entropy violates known science (a maximum entropy system has no available energy). Scientific-sounding statements like *"Since 'nothing' is as simple as it gets, we cannot expect it to be very stable. It would likely undergo a spontaneous phase transition to something more stable, like a universe containing matter"* [Ste07p133], and *"the three great conservation laws are not part of any structure. Rather they follow from the very lack of structure"* [Ste07p131], and constants *"must have changed very rapidly during the first moments of the big bang"* [Ste07p147] are foreign to known science. It should be noted that the largest spontaneous energy equivalent inferred by experiment is sub-atomic (and never on a vacuum with verifiably zero energy input, since even the solar neutrino flux is estimated to be $6 \times 10^{10}/cm^2/sec$ [Cosmic]).

"String Theory" has been the basis of several recently proposed multi-verse models. In these models, there are countless other universes in addition to our Universe (the contradiction of the "uni-" is ignored in these models). The building blocks for all of the universes are strings of vibrating energy in at least 10 dimensions (some models have 12 or more dimensions [Super, F-Theo]). It is speculated that our Universe arose out of a collision between previously existing universes, which arose out of collisions of their predecessors, and so forth. Each of these collisions produced matter, energy, and natural laws that are unique to that universe. The natural laws evolved during and shortly after the collision as the 10+ dimensions collapsed to four dimensions for our Universe. These are faith-based models based on unseeable 10-12 dimensions and innumerable unseeable universes. Nima Arkani-Hamed (Harvard) and others propose over 10^{500} universes because fewer would make the fine-tuning that is evident clearly infeasible [AAAS05]. Note that "String Theory" is not a scientific theory since it cannot be observed, tested, or falsified. *"The trouble is, proponents have not produced an iota of empirical evidence for strings. That's why University of Toronto physicist Amanda Peet – a proponent – recently called string theory a 'faith-based initiative'"* [Rev05]. *"No part of it has been proven, and no one knows how to prove it"* [Smo07].

Some have speculated that our Universe is a simulation designed by some advanced society in another universe [Bos02]. *"Another possibility is an unknown agent intervened in the evolution, and for reasons of its own restarted the universe in the state of low entropy characterizing inflation"* [Dys02]. *"To create a new universe would require a machine only slightly more powerful than the LHC* [Large Hadron Collider]... *The big question is whether that has already happened – is our universe a designer universe? By this, I do not mean a God figure,... there is still scope for an intelligent designer of universes as a whole... If our universe was made by a technologically advanced civilisation in another part of the multiverse, the designer may have been responsible for the Big Bang, but nothing more. ... intelligent designers create enough [$>10^{500}$] universes suitable for evolution, which bud off their own universes... It therefore becomes overwhelmingly likely that any given universe, our own included, would be designed rather than 'natural'"* [Gri10]. One can legitimately wonder what is meant by "overwhelmingly likely" since that implies that it is probable, which is not and cannot be demonstrated. In fact, the results from LHC have so far been very disappointing to many, failing to find evidence of the Higgs boson or supersymmetry (like "squarks" and "gluinos") [Web11].

The multi-universe speculations are meant to promote belief that such speculations are possible, and ultimately that they "must" be true. Since these universes would be outside our observability, even if they were true, the assertion that they are "possible" is unscientific since non-zero probability cannot be shown. *"Alternative universes, things we can't see because they are beyond our horizons, are in principle unfalsifiable and therefore metaphysical"* [Gef05].

"Because our Universe is, almost by definition, everything we can observe, there are no apparent mea-

surements that would confirm whether we exist within a cosmic landscape of multiple universes, or if ours is the only one. And because we can't falsify the idea, ... it isn't science" [Bru06]. *"If ... the landscape turns out to be inconsistent ... as things stand we will be in a very awkward position[,] without any explanations of nature's fine-tunings"* [Gef05]. It is clear that multi-verse is not science, but a philosophical belief.

Other models are largely ignored by cosmologists. John Wheeler *"ponders the question whether we humans actually create the laws by our observations, in the way that a magician creates illusion—that what we observe around us is no more real than what we observe at a magic show"* [Pri06]. In the Steady State Theory of Cosmology, new matter is continuously created as the Universe expands [Hoy93, Hoy95]. Plasma cosmology attributes the development of the visible Universe to interaction of electromagnetic forces on astrophysical plasma [Alf90]. The original ambiplasma was an equal mixture of ionized matter and anti-matter that would naturally separate as annihilation reactions released energy. Long-held models by most theistic religions typically involve "infinite" energy being(s) converting energy to mass or otherwise supernaturally creating the mass and energy of the Universe. These models, like the other origin models, cannot be tested or falsified by known science. Other models will undoubtedly arise, and as they do, each will need careful examination to verify any scientific validity.

Appendix B: Probability Estimates for Life-Support

The following estimates describe the probability that any planet (or comparable site) within the Universe will possess the specific features within the appropriate range to support bacterial life for 90 days or less. This list, prepared by astrophysicist Dr. Hugh Ross [Ros09], is based on a study of more than 650 research papers published in the astronomical and astrophysical literature. Even when the probabilities are adjusted for longevity and dependency factors, the odds in favor of even this simplest life form's brief existence in the observable Universe appear impossibly remote. How much more so the odds for life as advanced and enduring as human life!

Life-Support feature	Probability it will fall within the required range
relative abundances of different exotic mass particles.	.01
decay rates of different exotic mass particles.	.05
density of quasars in the local volume of the universe during early cosmic history.	.1
density of giant galaxies in the local volume of the universe during early cosmic history.	.03
galaxy size.	.01
galaxy type.	.1
galaxy mass distribution.	.02
size of galactic central bulge.	.05
galaxy location.	.01
variability of local dwarf galaxy absorption rate.	.2
quantity of galactic dust.	.2
giant star density in galaxy.	.2
star location relative to galactic center.	.2
star distance from corotation circle of galaxy.	.05
ratio of inner dark halo mass to stellar mass for galaxy.	.1
z-axis extremes of star's orbit.	.2
proximity of solar nebula to a normal type I supernova eruption.	.01
timing of solar nebula formation relative to a normal type I supernova eruption.	.01
proximity of solar nebula to a type II supernova eruption.	.01
timing of solar nebula formation relative to type II supernova eruption.	.01
timing of hypernovae eruptions.	.5

39

methane quantity in the atmosphere. 01
oxygen quantity in atmosphere. 1
nitrogen quantity in atmosphere. 001
carbon monoxide quantity in atmosphere. 1
chlorine quantity in atmosphere.. 1
cobalt quantity in crust and/or soil. 1
arsenic quantity in crust and/or soil. 1
copper quantity in crust and/or soil. 1
boron quantity in crust and/or soil. 1
cadmium quantity in crust and/or soil. 1
calcium quantity in crust and/or soil. 6
fluorine quantity in crust and/or soil. 1
iodine quantity in crust and/or soil. 2
magnesium in crust and/or soil. 4
manganese quantity in crust and/or soil. 1
nickel quantity in crust and/or soil. 1
phosphorus quantity in crust and/or soil. 02
potassium quantity in crust and/or soil. 4
tin quantity in crust and/or soil.. 1
zinc quantity in crust and/or soil. 1
molybdenum quantity in crust and/or soil. 05
vanadium quantity in crust and/or soil.. 1
chromium quantity in crust and/or soil.. 1
selenium quantity in crust and/or soil. 1
iron quantity in oceans.. 01
tropospheric ozone quantity. 2
stratospheric ozone quantity. 2
mesospheric ozone quantity. 2
quantity of greenhouse gases in atmosphere. 01
quantity of sea salt aerosols in troposphere. 2
ratio of electrically conducting inner core radius to radius of the adjacent
    turbulent fluid shell.. 2
ratio of core to shell (see above) magnetic diffusivity. .2
magnetic Reynold's number of the shell (see above). .2
elasticity of iron in the inner core.. 2
electromagnetic Maxwell shear stresses in the inner core. 2
core precession frequency for planet. 1
rate of interior heat loss for planet.. 1
quantity of sulfur in the planet's core.. 1
quantity of silicon in the planet's core. 1
viscosity at Earth core boundaries.. 01
viscosity of lithosphere. 2
thickness of mid-mantle boundary.. 1
intensity of primordial cosmic superwinds.. 05
number of smoking quasars. 05
formation of large terrestrial planet in the presence of two or more gas giant
    planets.. 01
total mass of Oort Cloud objects.. 3
mass distribution of Oort Cloud objects.. 3

hydrothermal alteration of ancient oceanic basalts. 1
location of dislocation creep relative to diffusion creep in and near the
    crust-mantle boundary (determines mantle convection dynamics). 1
number & mass of planets in system suffering significant drift. 01
mass of the galaxy's central black hole. 1
date for the formation of the galaxy's central black hole. 2
timing of the growth of the galaxy's central black hole. 4
rate of in-spiraling gas into galaxy's central black hole during life epoch. . . .05
distance from nearest giant galaxy. 5
distance from nearest Seyfert galaxy. 9
quantity of magnetars (proto-neutron stars with very strong magnetic fields)
    produced during galaxy's history. 3
ratio of galaxy's dark halo mass to its baryonic mass. 2
ratio of galaxy's dark halo mass to its dark halo core mass. 2
galaxy cluster formation rate. 1
tidal heating from neighboring galaxies. 5
tidal heating from dark galactic and galaxy cluster halos. 5
intensity and duration of galactic winds. 3
density of dwarf galaxies in vicinity of home galaxy. 1
amount of photoevaporation during planetary formation from parent star
    and other nearby stars. 1
in-spiral rate of stars into black holes within parent galaxy. 7
strength of magnetocentrifugally launched wind of parent star during its
    protostar era. 2
degree to which the atmospheric composition of the planet departs from
    thermodynamic equilibrium. 1
delivery rate of volatiles to planet from asteroid-comet belts during epoch
    of planet formation. 05
Q-value (rigidity) of planet during its early history. .2
injection efficiency of shock wave material from nearby supernovae into
    collapsing molecular cloud that forms star and planetary system. 01
number of giant galaxies in galaxy cluster. 1
number of large galaxies in galaxy cluster. 1
number of dwarf galaxies in galaxy cluster. 1
number and sizes of planets and planetesimals consumed by star. 3
distance of galaxy's corotation circle from center of galaxy. 1
rate of diffusion of heavy elements from galactic center out to the galaxy's
    corotation circle. 2
outward migration of star relative to galactic center. 3
viscosity gradient in protoplanetary disk. 1
average quantity of gas infused into the universe's first star clusters that
    reside in the vicinity of the potential life support galaxy. 1
level of supersonic turbulence in the vicinity of the potential life support
    galaxy during the infancy of the universe. 05
number and sizes of intergalactic hydrogen gas clouds in galaxy's vicinity. . .4
average longevity of intergalactic hydrogen gas clouds in galaxy's vicinity. . .4
avoidance of apsidal phase locking in the orbits of planets in the planetary
    system. 03
number density of the first metal-free stars to form in the vicinity of the

42

43

amount of methane generated in upper mantle of planet. 03
amount of buildup of heavy elements in the galaxy. 03
timescale for the buildup of heavy elements in the galaxy. 02
planet's silicate abundance. 1
timing of the 1:2 resonance event for Jupiter and Saturn. 1
intensity of superwinds generated by primordial supermassive black holes. .03
number of superwind events generated by primordial supermassive
    black holes. 03
mass of moon orbiting life support planet. 2
galaxy mass. 02
density of galaxies in the local volume around life-support galaxy   .1
average galaxy mass in the local volume around life-support galaxy. 1
rate at which the triple-alpha process (combining of three helium nuclei to
    make one carbon nucleus) runs inside the nuclear furnaces of stars. 002
average mass of cold dark gas-dust clouds in the galaxy. 3
number density of cold dark gas-dust clouds in the galaxy. 3
proximity of cold dark gas-dust clouds to life-support planet 1
masses of nearest cold dark gas-dust clouds to life support planet. 1
time in galactic history when cold dark gas-dust clouds form 3
intensity of the late heavy bombardment. 02
chemical composition of the late heavy bombarders. 1
depth of Earth's primordial ocean. 01
upper mantle seismic anisotropy. 1
lower mantle seismic anisotropy. 1
ratio of baryons in galaxy clusters to baryons in between galaxy clusters
    within the Local Volume of the universe. 1
ratio of baryons in galaxies to baryons in between galaxies in the Local
    Volume of the universe. 1
degree of central concentration of light-emitting ordinary matter for the
    life-support galaxy. 05
degree of flatness for the light-emitting ordinary matter for the life-support
    galaxy. 05
degree of sphericity for the distribution of ordinary dark matter for the
    life-support galaxy. 1
degree of sphericity for the distribution of exotic dark matter for the
    life-support galaxy. 1
average albedo of Earth's surface life. 01
level of carbon abundance in the galaxy. 05
gradient of carbon abundance with respect to distance from galactic center. .05
level of oxygen abundance in the galaxy. 05
gradient of oxygen abundance with respect to distance from galactic center. .05
level of nitrogen abundance in the galaxy. 1
gradient of nitrogen abundance with respect to distance from galactic center..1
infall velocity of galaxy toward center of nearest grouping of galaxies. 05
infall velocity of galaxy toward center of nearest supercluster of galaxies.. . .1
distance that primordial supernovae dispersed elements heavier than
    helium. 03
velocity of planet colliding with primordial Earth relative to Earth. 002
collision angle relative to Earth of planet colliding with primordial Earth. . . .05

vicinity of and inside the primordial Milky Way Galaxy. .... .04
solar system's orbital radius about the center of the Milky Way Galaxy. .... .01
quantity of soluble zinc in the oceans. .... .05
quantity of soluble silicon and silica in the oceans.. .... .05
quantity of phosphorous and phosphates in the oceans. .... .01
availability of light to upper layers of the oceans. 1
proximity of emerging solar system nebula to red giant stars .... .05
number of red giant stars in close proximity to emerging solar system
    nebula.. 1
masses of red giant stars in close proximity to emerging solar system
    nebula.. 1
proximity of emerging solar system nebula to fluorine-ejecting planetary
    nebulae.. .... .05
number of fluorine-ejecting planetary nebulae in close proximity to
    emerging solar system nebula. 1
number of large galaxy collisions with the Milky Way Galaxy during the past
    ten billion years. .... .03
    number of large galaxy collisions in the near vicinity of the Milky Way
    Galaxy during past ten billion years............................. .05
frequency of core collapse supernovae............................. .1
shape of the Milky Way Galaxy's ordinary dark matter halo. 1
mass of the potential life support planet.. .... .002
timing of potential life-support planet's birth relative to spiral substructure
    formation.. 2
luminosity variability of the primordial sun. 1
level of turbulence in the sun's primordial planetary disk.. 1
level of warping in the Milky Way Galaxy's spiral disk.................. .1
density of the galaxy.. .... .01
impact energy of moon-forming collidor event. .... .0001
Earth formation date relative to the formation date for the solar system
    nebula.. .... .02
flux of interplanetary dust into atmosphere. 7
density of particulates in the atmosphere.. 1
degree of suppression of dwarf galaxy formation by cosmic reionization in
    the local volume of the universe.. .... .02
silicon abundance in planetary system's primordial nebula.............. .01
rate of decrease of the thickness of the gas disk in the life-support galaxy.. .1
hydrogen escape from the atmosphere to outer space. .... .01
oxygen abundance in the galactic bulge. 1
production of H_3^+ by the galaxy's population III (first generation) stars. ..... .05
production of H_3^+ by the galaxy's population II (second generation) stars.. ... .05
intensity of ultraviolet radiation arriving from the sun at the time and shortly
    after life's origin on Earth (before photosynthesis can establish a
    significant ozone shield).. .... .002
wavelength response pattern of ultraviolet radiation arriving from the sun at
    the time or shortly after life's origin on Earth .... .02
gas density of the local interstellar medium.. .... .05
degree of oxidation of the phosphorus compounds in the protoplanetary
    disk of the solar nebula.. .... .05

mass of the disk of dust, asteroids, and comets for the primordial planetary system... .01
degree to which the solar wind penetrates Earth's magnetosphere......... .3
ratio of viscous to rotational forces in the planet's liquid core .... .01
inward migration of pebble-sized and smaller icy rubble from the outer primordial planetary disk....................................... .01
ratio of iron to chondritic meteorites at the time and place Earth's birth..... .01
number of ultracompact dwarf galaxies in the vicinity of the potential life support galaxy during that galaxy's youth.......................... .1
number of starless hydrogen gas clouds in the near vicinity of the potential life support galaxy.. .3
average mass of starless hydrogen gas clouds in the near vicinity of the potential life support galaxy...................................... .3
dust to gas ratio in and near the core of the potential life support galaxy during that galaxy's youth...................................... .1
dust temperature in and near the core of the potential life support galaxy during that galaxy's youth...................................... .1
gas temperature in and near the core of the potential life support galaxy during that galaxy's youth...................................... .1
dust to gas ratio in the mid to outer parts of the potential life support galaxy during that galaxy's youth................................. .1
dust temperature in the mid to outer parts of the potential life support galaxy during that galaxy's youth................................. .1
gas temperature in the mid to outer parts of the potential life support galaxy during that galaxy's youth................................. .1
quantity of carbon monoxide in the potential life support galaxy early in its history.. .1
number density of dark matter minihalos in the primordial Local Group..... .01
intensity or speed of high-velocity galactic outflows during the youth of the potential life support galaxy.................................. .01
thickness of the thick disk for the potential life support galaxy............ .03
epoch of peak production of type I supernovae in the potential life support galaxy... .1
average frequency of the different kinds of type I supernovae in the potential life support galaxy................................... .1
epoch of peak production of type II supernovae in the potential life support galaxy... .1
average frequency of the different kinds of type II supernovae in the potential life support galaxy................................... .1
virial radius of the exotic matter halo surrounding the potential life support galaxy... .02
mass of the corona surrounding the potential life support galaxy........... .1
diameter of the corona surrounding the potential life support galaxy........ .1
average strength of local gravitational instabilities in the potential life support galaxy.. .03
level of magnetic turbulence in the galactic interstellar medium............ .1
thermal pressure of the planet's ionosphere........................... .01
quantity of phosphorus mononitride and carbon monophosphide in the gas-dust cloud from which the solar system formed.................. .03

superwinds of large galaxies in the vicinity of the potential life support
   galaxy during the first two billion years of cosmic history............. .03
average size of cosmic voids in the vicinity of the potential life support
   galaxy.. .5
number of cosmic voids per unit of cosmic space in the vicinity of the
   potential life support galaxy................................... .5
number of galaxies per unit of dark matter halo virial mass in the vicinity
   of the potential life support galaxy.............................. .1
ratio of the number density of dark matter subhalos to the number density
   of dark matter halos in the vicinity of the potential life support galaxy..... .1
quantity of diffuse, large-grained intergalactic dust in the vicinity of the
   potential life support galaxy................................... .1
ratio of baryonic matter to exotic matter in dwarf galaxies in the vicinity of the
   potential life support galaxy................................... .1
ratio of baryons in the intergalactic medium relative to baryons in the
   circumgalactic medium for the potential life support galaxy 1
intergalactic photon density in the vicinity of the potential life support galaxy. .4
quantity of baryons in the warm-hot intergalactic medium in the vicinity of the
   potential life support galaxy................................... .2
radiative thermal conductivity level of the lower mantle................. .01
abundance of olivine in the upper mantle........................... .1
level of chemical heterogeneities throughout the lower mantle........... .1
rate at which the planet's inner core rotates faster than the mantle and
   the crust. 1
radiative thermal conductivity of the lower mantle.................... .01
quantity of Trichodesmium bacteria in the oceans.................... .0001
level of mixing in the early protoplanetary disk of the solar nebula. .... .05
distance of the Magellanic Clouds from the Milky Way Galaxy........... .5
timing of the movement of the main asteroid belt from its place of birth (much
   closer to the sun) to its present location (between Mars and Jupiter). 1

When the 10^{-614} probability for occurrence of all 501
parameters is corrected for dependencies and longevity,
the probability for occurrence of all 501 parameters
becomes 10^{-333}. The maximum possible number of life-
support bodies in observable Universe is 10^{22}. Thus, less
than 1 chance in 10^{311} exists that even one such life-support body would occur anywhere in the Universe for
bacteria to exist for 3 months. Long-term simple life and
intelligent life have much lower probabilities.

References (by primary author's last name, if any)

AAAS-4/11/05, "Harvard's Nima Arkani-Hamed Ponders New Universes, Different Dimensions," www.aaas.org/news/releases/2005/0511string.shtml

Abbot (Larry) , "The Mystery of the Cosmological Constant," Scient. Amer:(3), 1991.

Abel (David), "The 'Cybernetic Cut': Progressing from Description to Prescription in Systems Theory," Cybernetics and Systemics Journal (2), 2008, p252-262.

Abel (David), "The Biosemiosis of Prescriptive Information," Semiotica, 1/4/09, p1-19.

Abel (David), "The Universal Plausibility Metric (UPM) & Principle (UPP)," Theoretical Biology & Medical Modelling, 12/3/09, 6:27.

Alfven (Hannes) , "Cosmology in the Plasma Universe - an Introductory Exposition," IEEE Trans on Plasma Science: 18, 2/90, p5-10.

Appeals Court, 7th Circuit, Kaufman, James v. McCaughtry & Gary, 8/20/05.

Appeals Court, Comer v. TEA, 5th Circuit, 7/2/10.

Asimov (Isaac), "Can Decreasing Entropy Exist in the Universe?," Sci. Digest, 5/73, p76-77.

Astrobiology Magazine, "Inevitability Beyond Billions," 7/03.

Astrophysics Spectator: 2.33, 10/5/05.

Barbieri (Marcello), The Organic Codes. An Introduction to Semantic Biology, 2003.

Barbieri (Marcello), "Biosemiotics: a New Understanding of Life," Naturwissenschaften (95), 2/19/08, p577-599.

Barrow (John) & Joseph Silk, "The Structure of the Early Universe," Scientific American, 4/80, p118-128.

Bergman (Jerry), Slaughter of the Dissidents: The Shocking Truth about Killing the Careers of Darwin Doubters, 2008.

Bostrom (Nick), Anthropic Bias: Observation Selection Effects in Science and Philosophy, 2002.

Bowman (Lee), "Open Inquiry: the New Science Standard," Uncommon Descent, 10/20/07, www.uncommondescent.com/education/open-inquiry -the-new-science-standard/

Brady (Ronald), "On the Independence of Systematics," Cladistics: 1, 1985, p113-126.

Brumfiel (Geoff), "Outrageous Fortune," Nature, 1/5/06, p10-12.

Calvert (John), "Kitzmiller's Error: Defining Religion Exclusively rather than Inclusively," Liberty University Law Review: 3(2), Spr 2009, p213-328.

CosmicRays.org, "Solar Neutrinos," www.cosmicrays.org/muon-solar-neutrinos.php

Crocker (Caroline), Free To Think, 2010.

Davies (Paul), Superforce: The Search for a Grand Unified Theory, 1984, p235-236.

Davies (Paul), The Cosmic Blueprint: New Discoveries in Nature's Creative Ability To Order the Universe, 1988, p203.

Davies (Paul), Wiki-Quote,http://en.wikiquote.org/wiki/Darwinism

Dawkins (Richard),"The Necessity of Darwinism," New Scientist: 94, 4/15/82, p130.

Dawkins (Richard), The Blind Watchmaker, 1996.

Dawkins (Richard), A Devil's Chaplain: Reflections on Hope, Lies, Science, and Love, 2001, p79.

Denton (Michael), Nature's Destiny: How the Laws of Biology Reveal Purpose in the Universe, 1998.

Deutsch (David), Interviewed on The Science Show: The Anthropic Universe, 2/18/06.

Dyson (Lisa), Matthew Kleban, & Leonard Susskind, "Disturbing Implications of a Cosmological Constant," JHEP, 2/10/02, p11.

Eddington (Arthur), Space, Time and Gravitation: An Outline of the General Relativity Theory, 1920.

Expelled, The Movie, 2008.

F-theory, http://en.wikipedia.org/wiki/F-theory

Gefter (Amanda), "Is String Theory in Trouble?," New Scientist, 12/17/05.

GMIS (US Energy Department Genome Management Information System), http://genomics.energy.gov.

Grassé (Pierre-P), Evolution of Living Organisms, 1977.

Gribbin (John), Are We Living in a Designer Universe," UK Telegraph, 8/31/10.

Grossman(L), "Water's Quantum Weirdness Makes Life Possible," N Scient, 10/25/11.

Hawking (Stephen), A Brief History of Time, 1988.p205.

Hawking (Stephen) & Leonard Mlodinow, The Grand Design, 2010.

Hoyle (Fred), "The Universe: Past and Present Reflections," Engineering and Science, 11/81U, p8-12.

Hoyle (F.) & C. Wickramasinghe, Evolution from Space, 1981E.

Hoyle (F.), G. Burbidge, & J. Narlikar, Astrophysical Journal: 1-410 (2), 06/93, p437-457.

Hoyle (F.), G Burbidge & J Narlikar, "The Basic Theory Underlying the Quasi-Steady State Cosmological Model," Proc. R. Soc. A: 448, 1995, p191.

Hubble (Edwin), "A Relation Between Distance and Radial Velocity Among Extra-galactic Nebulae," Proc National Academy of Sciences: 15, 1929, p168–173.

Johnson (Donald), "Data and Information: Effect of Bioinformatics on Traditional Biology" (poster) Int Conf on Bioinformatics, 12/04.

Johnson (Donald), Programming of Life, 2010.

Johnson (Phillip), "The Church of Darwin," Wall Str Jour, 8/16/99.

Kormendy (J.), D. Richstone, "Inward Bound---The Search For Supermassive Black Holes In Galactic Nuclei," Annual Reviews of Astronomy and Astrophysics: 33, 1995, p581-624.

Lewontin (Richard), "Billions and Billions of Demons," in NY Review of Books, 1/9/97.

Luskin (Casey), "Does Challenging Darwin Create Constitutional Jeopardy? A Comprehensive Survey of Case Law Regarding the Teaching of Biological Origins," Hamline University Law Review: 32(1), 2009, p1-64).

Mazur (Suzan), The Altenberg 16: An Exposé of the Evolution Industry, 2010.

Michael (Eli), "How Physically Plausible is the Cosmological Constant?," U-CO, 1999, super.colorado.edu/~michaele /Lambda/phys.html

Morris (M.), K. Thorne, & U. Yurtsever, "Wormholes, Time Machines, and the Weak Energy Condition," Physical Review: 61 (13), 9/88, p1446-1449.

NASA, On the Expansion of the Universe, www.grc.nasa.gov/WWW/K-12 /Numbers /Math/documents/ON_the_EXPANSION _of_the_UNIVERSE .pdf

Occam's razor, en.wikipedia.org/wiki/Occam%27s_Rasor

OOLprize, Origin of Life Prize, www.us.net/life/

Open Letter to Scientific Community, New Scientist, 5/22/04, cosmologystatement.org

Overbye (Dennis), "Dark energy is Still Puzzle to Scientists," NY Times, 6/4/08.

Pagels (H. R.), The Dreams of Reason, 1988, p156-58.

Pauling (L), Pauling Institute, lpi.oregonstate.edu/lpbio/lpbio2.html

Peebles (P. J. E.), & Bharat Ratra, "The Cosmological Constant and Dark Energy," Reviews of Modern Physics: 75, 2003, p559–606..

Penrose (Roger), The Emperor's New Mind: Concerning Computers, Minds, and the Laws of Physics, 1989, p344.

Polanyi (Michael), Quote originally at Michael Polanyi Center site, now at nostalgia .wikipedia.org/wiki /Michael_Polanyi

Popper (Karl), "Science as Falsification,"Conjectures and Refutations, 1963, p33-39.

Pound (R. V.) & J. L. Snider, "Effect of Gravity on Gamma Radiation," Phys. Rev.: 140(3B), 1965, B788-803.

Princeton Physics News: 2 (1), 2006, p6.

Provine (Will), "No Free Will," in Catching Up with the Refutations, 1963, p33-39.

Rusbult (Craig), "Methodological Naturalism in Our Search for Truth: A Brief Introduction," www.asa3.org/asa/education/origins/briefmn.htm

Ross (Hugh), "RTB Design Compendium (2009)," Part 3; accessed 6/14/11, http://www.reasons.org/files/compendium /compendium_Part3 _ver2 .pdf.

Sagan (Carl), "Can We Know the Universe?," in Broca's Brain, 1979, p13-18.

Sewell (Granville), "Can 'ANYTHING' Happen in an Open System?," in The Numerical Solution of Ordinary and Partial Differential Equations, 2005, Appendix D.

Smolin (Lee), Life of the Cosmos, 1997, p44-45.

Smolin (Lee), The Trouble with Physics, 2007.

Stenger (Victor) , God: The Failed Hypothesis: How Science Shows That God Does Not Exist, 2007.

Superstrings, en.wikipedia.org/wiki/Superstring_theory

Supreme Court Decision, Torcaso v. Watkins (367 U.S. 488), 1961.

Supreme Court Decision, US v. Seeger, 380 U.S. 163, 1965

Supreme Court Decision, Gillette v. U.S., 401 U.S. 437, 450, 1971.

Supreme Court Decision, Harris v. McRae, 448 U.S. 297, 1980.

Swinburne (Richard), "Argument From the Fine-Tuning of the Universe," in Physical Cosmology and Philosophy, 1990, p154-73.

Trevors (J. T.) & D. L. Abel, "Chance and Necessity Do Not Explain the Origin of Life," Cell Biology International: 28, 2004, p729-739.

Turner (Michael), "Hawkings: No miracle in the multiverse," Nature:467, 10/7/10, p657-658.

Venter (Craig), Interview,http://www.guardian.co.uk/science/video/2010/may/20 /craig-venter-new-life-form

Wald (George), "The Origin of Life" in The Physics and Chemistry of Life, 1955, p12.

Webb (Richard), "Should We Worry about What the LHC Is Not Finding?," New Scientist, 7/25/11.

Webster's Third New International Dictionary of the English Language, 1993.

Weinberg (Steven), "Life in the Quantum Universe," Articles: Scientific American, proxy.arts.uci.edu/~nideffer /Hawking/early_proto/weinberg.html

Wilson.(D), "Atheism as a Stealth Religion," Huff. Post, 12/14/07

Wolfenden (Richard), in Without Enzyme Catalyst, Slowest Known Biological Reaction Takes 1 Trillion Years, 2003, www.unc.edu/news/archives/may03/enzyme050503.html.

Young (Willard), Fallacies of Creationism, 1985, p165.